Fausto Batista Felix Silva

Proposal to automate the capture of solar rays in PV systems

AF537052

Fausto Batista Felix Silva

Proposal to automate the capture of solar rays in PV systems

Application in solar photovoltaic systems

ScienciaScripts

Imprint

Any brand names and product names mentioned in this book are subject to trademark, brand or patent protection and are trademarks or registered trademarks of their respective holders. The use of brand names, product names, common names, trade names, product descriptions etc. even without a particular marking in this work is in no way to be construed to mean that such names may be regarded as unrestricted in respect of trademark and brand protection legislation and could thus be used by anyone.

Cover image: www.ingimage.com

This book is a translation from the original published under ISBN 978-613-9-66528-0.

Publisher:
Sciencia Scripts
is a trademark of
Dodo Books Indian Ocean Ltd. and OmniScriptum S.R.L publishing group

120 High Road, East Finchley, London, N2 9ED, United Kingdom
Str. Armeneasca 28/1, office 1, Chisinau MD-2012, Republic of Moldova, Europe
Printed at: see last page
ISBN: 978-620-8-03625-6

Copyright © Fausto Batista Felix Silva
Copyright © 2024 Dodo Books Indian Ocean Ltd. and OmniScriptum S.R.L publishing group

It's fascinating to look at the past and see how many sacrifices, how many efforts, how many worries and how many obstacles we've overcome, but it's even more fascinating to look to the future with faith, knowing that there is only one God, who always accompanies me showing me that for him all things are possible, you just have to believe, so I dedicate this victory first of all to God for life and throughout my journey he has always been by my side carrying me in his arms and working great miracles.

SUMMARY

SILVA, Fausto B. F. **Proposal to automate the collection of solar rays for photovoltaic systems.** 2015. 52 f. Monograph (Specialisation Course in Industrial Automation), Academic Department of Electronics, Federal Technological University of Paraná. Curitiba, 2015

Due to the growing demand for electricity production around the world, it is necessary to think of alternatives to meet the need for this growth. Renewable energy sources are currently being developed in several countries, helping to increase energy production while also ensuring that these sources are clean, without damaging the environment, which means not emitting fossil fuels. One of the clean energy alternatives is photovoltaic energy, which transforms solar energy into electrical energy using solar panels. The purpose of this work is to propose a solution for automating the capture of the sun through solar panels, because once the panel is installed in a fixed way, it will not achieve optimum energy efficiency, taking into account that the sun changes its trajectory over the course of the days and the year. This work also presents research on the subject that shows the state of the art, so it is possible to propose improvements to existing systems. The aim of this work is not to design the project, but to carry out research to assess the feasibility of implementing automation and control of solar energy collection.

Key words: Solar energy. Photovoltaic panels. Automatic sun tracking.

SUMMARY

CHAPTER 1

INTRODUCTION

The following chapter presents the subject of the work, showing solar energy as a source of renewable energy, the delimitation of the study, the problem and justification relating to the subject, the objectives of the work, as well as the methodological procedures, its structure and the schedule of activities that will be carried out.

1.1 THEME

One of the planet's main sources of energy is the sun, which emits a large amount of energy in the form of light and heat, capable of supplying the entire world's need for this type of energy, often associated with renewable sources. A major advantage of using clean energy is that it generates practically no waste compared to fossil fuels, which generate a large amount of waste or pollutant emissions into the environment, and is an alternative to the exponential growth in demand for electricity around the world (VILLALVA; GAZOLI, 2012).

Solar energy can be harnessed in two ways: as photothermal energy, which consists of capturing solar radiation and converting it directly into heat in order to heat water using the sun's heat directly, without the need to use other resources. Another form of utilisation is photovoltaic solar energy, which consists of the direct conversion of light into electrical energy through the photo voltaic effect (VILLALVA; GAZOLI, 2012).

What differs between the two ways of using solar energy is that unlike photothermal systems, which are used for heating, photovoltaic systems are able to directly capture sunlight and produce a difference in potential through photovoltaic plates fixed to roofs and façades, processed by controllers and converters, which can be stored in batteries or used directly in systems connected to the electricity grid (VILLALVA; GAZOLI, 2012).

This study deals with the subject of capturing solar rays for photovoltaic systems. The research to be carried out converges on a review of the state of the art on the subject, pointing out the effects of capturing solar energy in photovoltaic modules fixed to roofs or façades, the best angle for capturing light and mapping the sun's trajectory.

1.2 DELIMITATION OF THE STUDY

According to Villalva and Gazoli, there is no consensus on the best method for choosing the best angle of inclination for a photovoltaic module, but it is possible to determine an angle of inclination for a given geographical latitude that allows for adequate average energy production throughout the year.

The study aims to analyse solutions for automating this process, looking for references of solutions applied to this problem and, within these references, pointing out improvements that can be made to improve energy capture and, as a consequence, improve the efficiency of this system.

1.3 PROBLEM

Currently, the use of photovoltaic solar energy in Brazil is used not only in homes, but also in small industries, rural areas, remote telecommunications centres and government programmes such as the Light for All programme (PORTAL BRASIL, 2014). In the future, photovoltaic solar energy will be concentrated in systems connected to the electricity grid (MOLGARO, 2014).

Today, the energy produced by this collection system is somewhat limited in terms of how much light each photovoltaic module is able to capture due to its fixed installation. Because the angle of inclination of the module is fixed, it is not possible to maximise the capture of the sun's rays on all days or months of the year or throughout the day, making it necessary to adopt an angle that allows an average production of energy in all seasons of the year, because the incorrect choice of inclination reduces the capture of the sun's rays and compromises the production of electricity by the photovoltaic module (VILLALVA; GAZOLI, 2012).

There is no general consensus on the best method for choosing the angle of inclination for installing a solar module, so: **How can you automate the collection of solar rays for photovoltaic systems and thus have better energy efficiency at all times of the year?**

It is believed that with a study of the state of the art and the proposals presented throughout this work, an improvement can be made to the automation system for capturing solar rays.

1.4 OBJECTIVES

This section presents the general and specific objectives of the work, relating to the problem presented above.

1.4.1 General Objective

The aim of this project is to present a proposal for automating the capture of solar rays through the best angle of solar incidence, increasing its energy efficiency by identifying the characteristics needed to automate this process.

1.4.2 Specific Objectives

- Compare the energy efficiency of photovoltaic panels that are installed at fixed points;
- Map the angle of solar declination on different days of the year;
- Compare the advantages and disadvantages of a solar energy collection system that follows the path of the sun;
- Identify the automation elements of this system in the state of the art.

1.5 BACKGROUND

Most photovoltaic systems are made up of plates with a fixed angle of inclination. With fixed-tilt plates, it is not possible to maximise the capture of the sun's rays on all days or months of the year, making it necessary to adopt an angle that allows a reasonable average energy production throughout the year and throughout the day, so the incorrect choice of tilt reduces the capture of the sun's rays and compromises the production of electricity as well as its energy efficiency (VILLALVA; GAZOLI, 2012).

An automation system for capturing the sun's rays for photovoltaic systems or automatically tracking the sun's position can optimise the angle of incidence of the sun's rays automatically throughout the day and throughout the months of the year, taking into account that due to the existence of the solar declination angle, the sun rises and sets at different points in the sky and describes a trajectory with a different inclination each day of the year.

Automating the process will increase energy capture in the solar modules by tracking the movement of the sun, thus making them more energy efficient, as the photovoltaic panels will always receive the sun's rays at the best possible angle of incidence.

1.6 METHODOLOGICAL PROCEDURES

In terms of research classification, this is research that seeks an answer to the problem that is the subject of this study, i.e. applied research and the development of this work will consist of a number of stages involving: the study to define solutions or characteristics for automating the process of capturing solar energy, the measurement of energy generated by fixed plates, mapping the sun's trajectory on different days and months of the year.

The study of automation solutions will be based on existing literature, checking the most commonly used methods and thus proposing viable improvements for optimising solar energy capture, in the form of a bibliographical survey. For energy measurement by fixed plates, the aim is to measure the energy efficiency of this type of installation and then compare it with plates that follow the path of the sun. In order to map the sun's trajectory, the study will use *software*, since a complete analysis requires mapping of all the periods and seasons of the year.

Among the existing literature on the subject, there is a study carried out in Bucharest that presents two types of automatic electro-mechanical trackers. In the first study, the solar energy detectors are photovoltaic cells or thermal resistors. Two devices are connected in a differential circuit: the first device receives deviating radiation from the east towards the sun and the second device receives radiation from the west towards the sun. When the photovoltaic plates are orientated towards the sun, the detector signals do not detect an error signal, in all other situations an error signal causes the motor to follow the movement of the sun from a control system until the error is corrected. The second study shows a low-cost monitoring procedure using the amplitude-based comparison method and a special micro-signal detector made with Micro Electro Mechanical System (MEMS) technology. The procedure allows simple panel rotations based on complete tracking, by azimuth and solar height, from a predetermined adjustment angle in the range of 30 -150° (MILEA *et al,* 2007).

Another automatic tracking project presented at a symposium on excellence in management and technology shows the design and development of a solar azimuth and elevation tracker, based on a new conception of the positioning sensor, made up of four photoresistors that are independent of the solar coordinate relationship. It uses two direct current motors that act on the vertical and horizontal axes, controlled by a PIC microcontroller (RIBEIRO, PRADO, GONÇALVES, 2012).

1.7 THEORETICAL BACKGROUND

On the subject of solar diagrams and mapping, Lamberts (2012), on the automation characteristics of the system, MILEA *et al,* (2007), Ribeiro, Prado, Gonçalves (2012) and Trevelin (2014). As for solar positioning concepts, solar angles and theoretical foundations, Villalva and Gazoli (2012).

1.8 WORK STRUCTURE

The work will have the following structure.

Chapter 1 - Introduction: the topic, the study's delimitations, the problem, the research objectives, the justification, the methodological procedures, the indications for the theoretical basis, the general structure of the work and the schedule of activities to be carried out are presented.

Chapter 2 - Theoretical Basis: this covers the basics of operation, the different technologies and devices used to automate the system, as well as a background on the solar trajectory.

Chapter 3 - Development: shows the solar mapping to determine the solar trajectory and the characteristics for the automation proposal.

Chapter 4 - Results and discussions: the results obtained with the automatic solar capture tracking system are discussed, based on the specifications initially proposed.

Chapter 5 - Final considerations: the research question and objectives are revisited and it is pointed out how they were solved, answered or achieved through the work carried out.

In addition, future work that can be carried out on the basis of this study is suggested.

CHAPTER 2

THEORETICAL BACKGROUND

To begin the study on automating the capture of solar rays for photovoltaic systems, it was necessary to understand the potential of photovoltaic energy in Brazil, the fundamentals of how this system works, the different types of photovoltaic cells, elements for the automation process, the functioning and types of sensors, and an analysis of the sun's trajectory. These subjects will be dealt with in the course of this chapter.

2.1 THE POTENTIAL OF PHOTOVOLTAICS IN BRAZIL

With technological advances and industrial growth around the world, there has been an exponential increase in the production of electricity, which raises concerns about the way in which energy is produced, often causing damage to the environment. As a way of producing electricity in a clean way, photovoltaic power generation technology has been gaining more and more ground for use in vehicles and power stations in homes and small industries (Villullas *et al,* 2011).

Photovoltaics have long been used as a form of clean and sustainable energy production, based on the most abundant and widely available renewable energy source on the planet - the sun. Although Brazil has great potential for generating photovoltaic energy, only IGWh of energy is produced, while Europe has 88GWh of photovoltaic energy installed. Figure 1 compares the solar irradiation values for Brazil and Europe (Portal Solar, 2015).

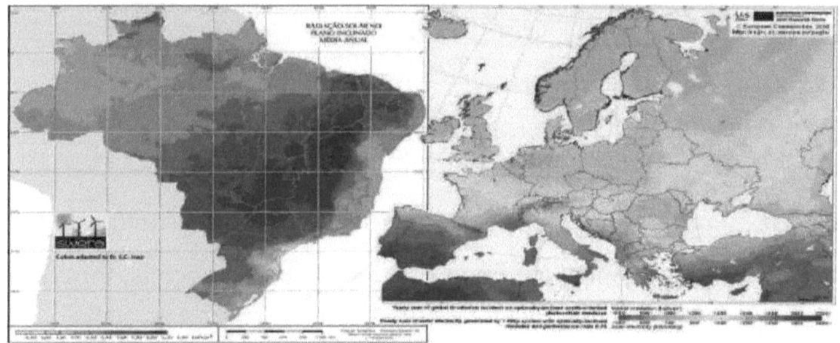

Figure 1 - Photovoltaic energy potential **Source:** Portal Solar (2015)

2. 2PHOTOVOLTAIC CELLS

The photovoltaic effect is a physical phenomenon in which sunlight is converted into electricity, which occurs when light or electromagnetic radiation from the sun, which is made up of photons with a large amount of energy, is incident on a cell made up of semiconductor materials. Although both materials are electrically neutral, N-type silicon has excess electrons (-) and P-type silicon has excess gaps (+). Interleaving these creates a P-N junction and an electric field. When these two semiconductors are interleaved, the excess electrons from the N-type flow into the P-type semiconductor, and the electrons that have left the N-type then create gaps in it. Through the flow of electrons and gaps, the two semiconductors act like a battery and create an electric field at the P-N junction. It is this field that makes the electrons available for the electrical circuit. At the same moment, the gaps move in the opposite direction, towards the positive surface where they await free electrons (Molgaro, 2014).Figure 2 shows the structure of a photovoltaic cell made up of a semiconductor material with P and N layers.

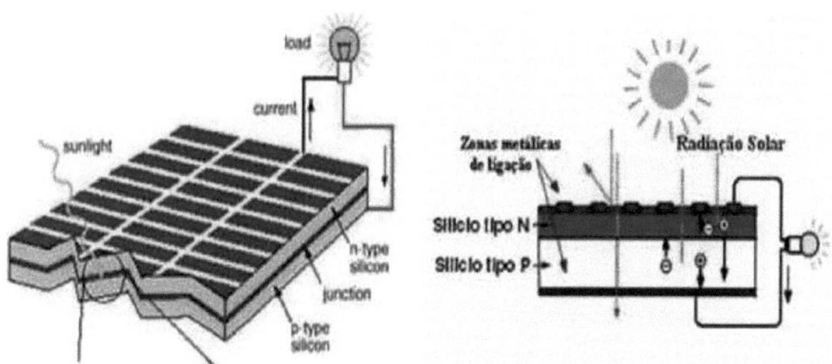

Figure 2 - Structure of a photovoltaic cell

Source: Viridian (2015)

2.3 TYPES OF PHOTOVOLTAIC CELLS

There are currently several technologies and materials for manufacturing photovoltaic cells and modules, the most common of which are silicon cells, thin film and also as

a solution to make cells made from organic material more versatile in terms of energy capture. The characteristics of these different technologies are presented below.

2.3.1 Monocrystalline Silicon Cell

Monocrystalline silicon cells are designed using high purity silicon blocks heated to high temperatures and subjected to a crystal formation process called the Czochralski method, which results in a product called monocrystalline silicon ingot, made up of a single crystalline structure with a homogeneous molecular organisation. The formed ingot is sliced to produce *wafers* and only then is it subjected to chemical processes, receiving impurities on both sides, thus forming the P and N layers and forming the basis for the operation of the photovoltaic cell. Only after these processes have been completed does the plate receive a metallic film on one side, a metallic grid on the other side and a layer of antireflective material that will receive the light, obtaining the final product of the monocrystalline silicon cell as shown in Figure 3 (Villalva, Gazoli, 2012).

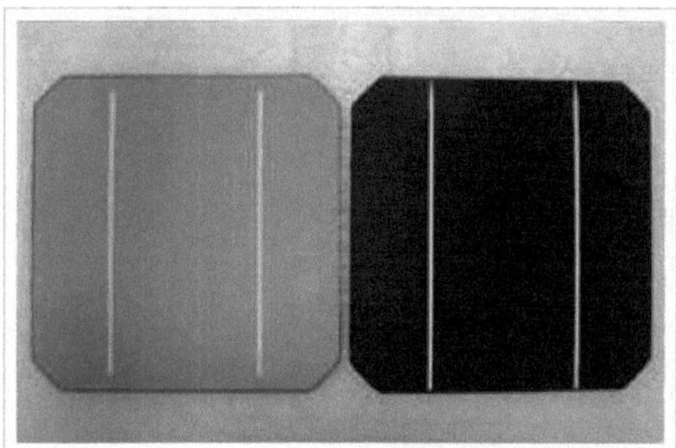

Figure 3 - Monocrystalline Silicon Cell
Source: Topsky (2011)

Although they have a higher production cost compared to other photovoltaic cells, monocrystalline silicon cells are the most efficient found commercially due to their better energy efficiency of between 15 and 18 per cent

2.3. 2Polycrystalline Silicon Cell

The production process for this cell is obtained by ingots, which are made up of a cluster of small crystals of different sizes and orientations. Just like monocrystalline silicon cells, these cells are also cut to produce *wafers* that will be transformed into photovoltaic cells whose predominant colour is blue, but can change depending on the antireflective material, as shown in Figure 4. The advantage of using a polycrystalline silicon cell is that it has a lower manufacturing cost than a monocrystalline cell, although it has a lower commercial efficiency of around 13-15% (Villalva, Gazoli, 2012).

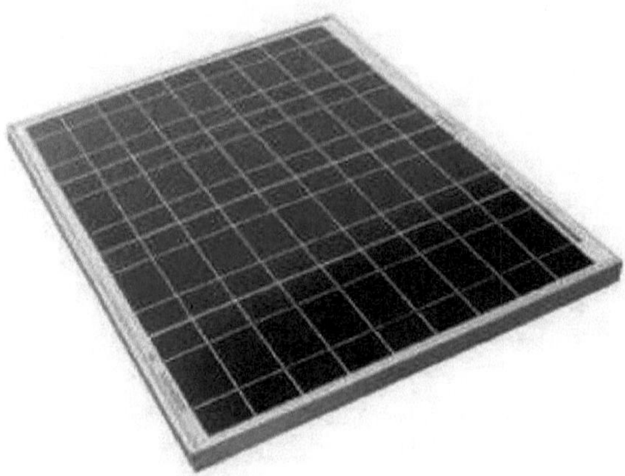

Figure 4 - Polycrystalline Silicon Cell
Source: Topsky (2011)

2.3.3 Thin Films

Recent technologies for producing photovoltaic cells include thin films, which, unlike technologies using crystal silicon, also use other materials that can be applied to a rigid or flexible base. The advantage of producing this type of cell is the small amount of raw material used and the deposition process, which can be by vaporisation or another method, thus reducing the waste that occurs in other methods, such as sawing to create *wafers* and also the

The manufacturing temperature is much lower, between 200 and 500°C, than the temperatures of up to 1500°C used in the manufacture of crystalline cells. As the manufacture of thin films is less complex and the raw materials consume little energy, their cost is also lower, making large-scale production possible in various sizes.

Although they cost less, their energy efficiency is also lower, requiring a larger module area to produce the same energy as crystalline silicon cells, which is not difficult since thin film cells make better use of sunlight (Villalva, Gazoli, 2012). Figure 5 shows a thin film cell on a flexible base.

Figure 5 - Thin Film Cell **Source:** Almaks (2015)

2.3.4 Amorphous Silicon

It was the first thin-film technology developed and has an energy efficiency of between 5 and 8 per cent, which is very low compared to crystalline cells where this percentage is even lower, because due to its light-induced degradation after 12 months its efficiency decreases until it becomes stable (Villalva, Gazoli, 2012).

2.3.5 Microcrystalline Silicon Thin Film

This technology combines the advantages of crystalline silicon and thin film manufacturing technology, which means it can be produced on a large scale, has a lower cost, less waste and a higher efficiency of around 8.5%. It is manufactured in two stages, one at high temperature and the other at low temperature (Villalva, Gazoli, 2012).

2.3.6 Cadmium telluride cells

Although cadmium telluride (CdTe) cells are the most efficient within the thin film family, reaching 14.40 per cent efficiency, they cannot be produced on a large scale because cadmium (Cd) is a toxic material and tellurium (Te) is difficult to find

(Villalva, Gazoli, 2012).

2.3.7 CIGS cells

The CIGS (Copper, Indium, Gallium and Selenide) cell is still little used commercially due to its high cost, although its construction does not use toxic materials and its energy efficiency is up to 14 per cent (Villalva, Gazoli, 2012).

2.3.8 Organic Solar Cells

Organic solar cells are made from carbon. Unlike silicon solar panels made on a rigid base, organic solar cells are produced on a flexible plastic material, where the components are deposited using a printing technique, which is an advantage since this technique allows for large-scale production.

> "Carbon has the potential to offer high performance at low cost. From our best information, this is the first demonstration of a functional solar cell that has all its components made of carbon," explained Dr Zhenan Bao, from Stanford University (INEO,2015).

The organic solar cell is made up of a photoactive layer, which absorbs sunlight, applied between two electrodes. The active layer is made up of carbon nanotubes and fullerenes *(buckyballs)*. The electrodes in a typical organic solar cell are made of metals - gold, silver or copper - and indium tin oxide (ITO), the material used in touchscreens (INEO, 2015). The advantage of using this type of cell is that it costs less than silicon cells, has a high energy efficiency of around 21 per cent and, because they are manufactured by a printing process, can be installed more flexibly, as shown in Figure 6.

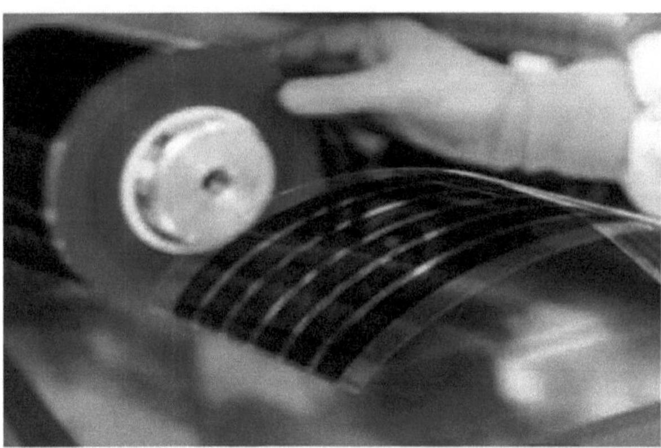

Figure 6 - Organic photovoltaic cell **Source: CSEM** (2013)

1.1.9 Comparing Different Technologies

Thin-film solar panels accounted for 11 per cent of all solar panel sales in 2011. Production capacity is expected to grow at an annual rate of 24 per cent, reaching more than 22 GW by 2020 (or a global market share of 38 per cent in terms of module production). There are three main types of thin-film solar panels on the market today: amorphous silicon (a-Si), cadmium telluride (CdTe) and copper indium gallium selenide (CIS / CIGS). Table 2 shows their main characteristics:

Table 1 - Comparison of different technologies

	a-Si	CdTe	CIGS
Better efficiency	13,40%	19,00%	20,40%
More efficient module	8,10%	14,40%	14,50%
Thin film market action	32,00%	43,00%	25,00%
Advantages	Excellent technology for small devices	Low production costs	High efficiency
Disadvantages	Low efficiency and expensive equipment	Medium efficiency, rigid glass and highly toxic cadmium components	Slow growth due to expensive market share of traditional process. Requires less cadmium than CdTe in solar cells
Main Manufacturers	Sharp	First Solar	Solar Frontier

Source: Energy Infonnative (2015)

2.4 STEP ENGINE

Although stepper motors are widely used in the computer industry, they have also been widely used outside of it, such as in military, medical, commercial and automation applications, because a stepper motor is a device that converts a digital input into a mechanical output.

Its construction consists of a slotted stator equipped with two or more individual coils and a rotor without windings. The motor is classified based on its construction, which involves the use or not of a permanent magnet, i.e. if the magnet is used it is classified as a permanent magnet stepper motor, if the magnet is not used it is classified as a reluctance stepper motor (TORO, 1994).

Its principle of operation consists of using two solenoids aligned two by two which, when energised, cause the rotor to be attracted to them, aligning it to the axis determined by the solenoid, causing a variation in the angle (Santos, 2008).

2.5 SOLAR TRAJECTORY

Because of the solar declination angle, the sun rises and sets at different points in the sky and describes a trajectory with a different inclination each day of the year. In the rotational movement, the earth rotates around an east-west axis characterised by the azimuthal angle, which passes through its poles, giving rise to day and night. The earth's movement in translation takes place on the north-south axis, determining the four different seasons of the year, for each of which the sun has a characteristic movement, such as solstice, which is the time when the sun passes through its greatest boreal or austral declination, and during which it ceases to move away from the equator, and also equinox, which is the point in the earth's orbit at which there is an equal length of day and night. These events occur according to the seasons, as shown in Table 1 (Lamberts, 2015).

Table 1 - Solar Characteristics

Date	Name
21st March	Autumn Equinox
21st September	Spring Equinox
21st June	Winter solstice
21st December	Summer solstice

Source: Lamberts 2015

2.5.1 Azimuth angle

The azimuthal angle or solar zimuth is the angle of orientation of the sun's rays in relation to geographic north and is described throughout the day, so if an observer is located in the southern hemisphere, depending on the time of day the sun will be in a different position, to the left or right of the observer and at midday the sun will be exactly in front of them, which means that the azimuthal angle is zero (Villalva, Gazoli, 2012).

For observers located in the northern hemisphere, the azimuth angle is measured in relation to the geographic south.

2.5. 2Solar altitude

Each day of the year the sun rises and sets at different points in the sky, describing paths with different inclinations, due to the existence of solar declination angles, as shown in Figure 7.

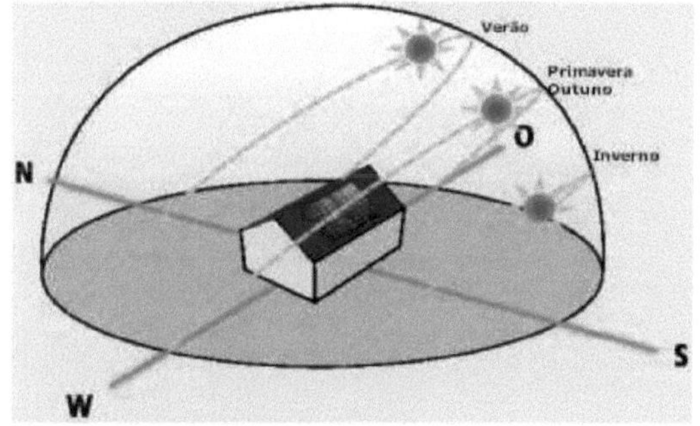

Figure 7 - Solar position and spectrum
Source: Electrónica (2015)

Looking northwards, on summer days the azimuthal angle is smaller, on winter days the solar height in the sky is lower, which means that the sun can be seen lower to the horizon, so the azimuthal angle and the air mass travelled by the sun's rays are greater than in summer. The angle of solar height (Ys) depends on the observer's geographical location and also on the angle of solar declination, i.e. if the observer is close to the equator, they will see greater solar heights and if they are close to the

Earth's poles they will see lower heights (Trevelin 2014).

However, in order to analyse the sun's trajectory, it is necessary to consider the earth's rotational movement on the east-west axis, characterised by the azimuthal angle (0_a) and the translational movement on the north-south axis, characterised by the solar height angle (Ys), as shown in Figure 8. This angular variation has an impact on the generation of electricity by a photovoltaic panel, often causing low energy efficiency depending on various factors such as time of day, month and even season (Trevelin 2014).

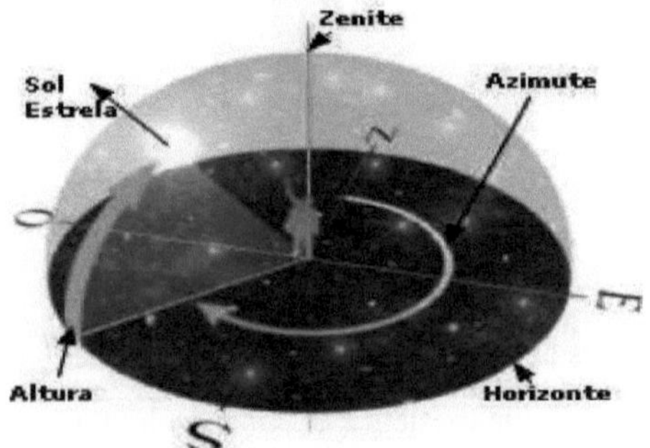

Figure 8 - Position of the sun in relation to the azimuthal, zenithal and solar height angles. **Source:** Electrónica (2015)

CHAPTER 3

DEVELOPMENT

In order to present a proposal for automating the collection of solar rays for photovoltaic systems, it is necessary to identify the characteristics that make up this automation, which will be presented throughout this chapter. Figure 9 shows the composition of the features for automation.

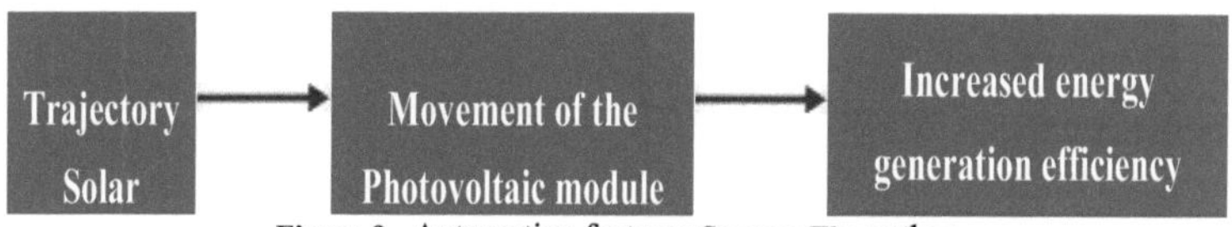

Figure 9 - Automation features **Source: The** author

3.1 SOLAR TRAJECTORY

As one of the aims of the automation proposal is to make the photovoltaic (PV) panels follow the path of the sun, the different paths of the sun will be presented below.

3.1.1 Solar trajectory during the day

It is possible to analyse during the day that the sun rises and sets in different places during the day, thus tracing different angles as it changes its trajectory throughout the day. Figure 10 shows the solar trajectory in Curitiba over the course of a day, where by analysing the figure you can see the positions of the sun at sunrise and sunset. The thinner curve indicates the sun's current trajectory, and the yellow area around it indicates the variation of the sun's trajectories during the year. The colours in the figure's time bar show the sunlight coverage during the day, indicating dawn at 06:3 Ih, sunrise at 06:5511, solar noon at 12:24h, sunset at 17:53 and dusk at 18:17.

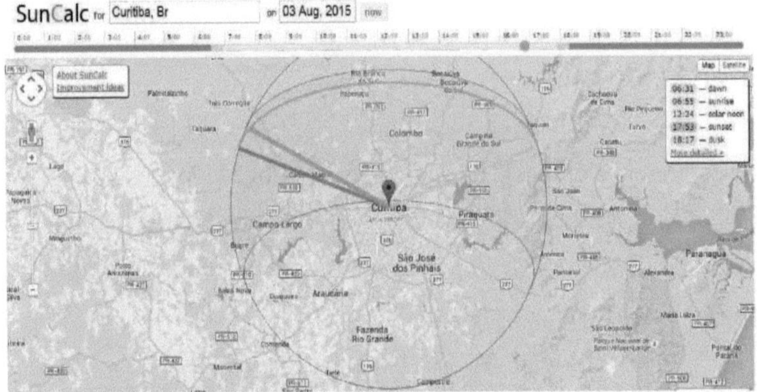

Figure 10 - Solar orientation during the day **Source:** SunCalc (2015)

3.1.2Solar trajectory during the year

The first step in determining the automatic control of the angle of the photovoltaic panels is to determine the solar trajectory throughout the year as a function of the city's latitude in order to visualise the sun's angle of inclination at different times. Figure 11 shows the solar chart for the city of Curitiba indicating a latitude of 25.51° and the variations that exist at different times of the year.

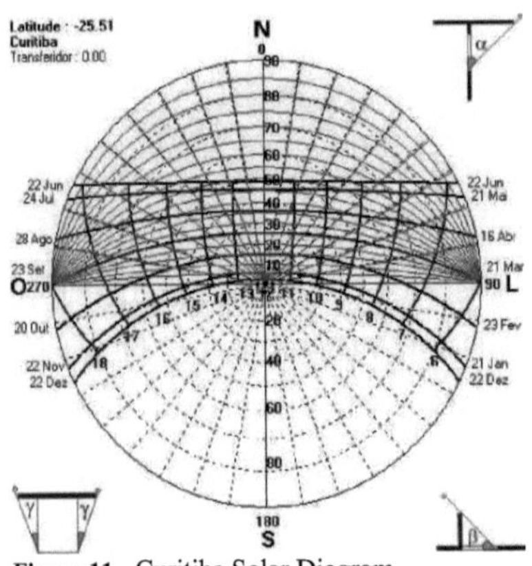

Figure 11 - Curitiba Solar Diagram
Source: SOL-AR 6.2 ***software***

To better visualise the solar trajectory in the city of Curitiba, Figure 12 shows a diagram of the sun's apparent movement at a latitude of 25.5° obtained from the previous figure.

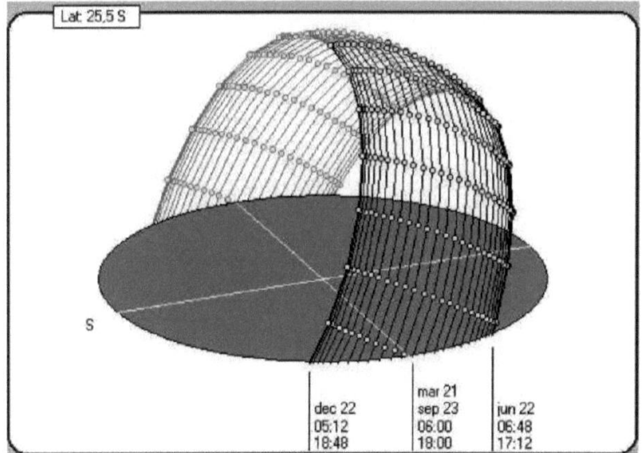

Figure 12 - Apparent movement of the sun
Source: ***Sunpath*** 1.0 ***software***

With the diagram and the determination of solar movement, it is possible to visualise the inclination of the sun's trajectory at different times of the year. This inclination directly influences the energy efficiency produced by the photovoltaic panels, since the panels are installed in a fixed way, which means that depending on the sun's movement, depending on the time of year, the panel will have greater or lesser efficiency.

3.2MOVEMENT OF THE PHOTOVOLTAIC MODULE

To determine the system for moving the photovoltaic modules, it is necessary to gather information on the characteristics that directly influence this process and that make automation viable, efficient and low in cost and energy consumption.

3.2.1 Degrees of Freedom

Solar modules that automate the capture of the sun's rays can have one or two degrees of freedom, with one being adjusted in relation to geographic north (azimuth angle) and the other in relation to the sun's altitude, i.e. with one degree of freedom it is possible to track the sun throughout the day and with an additional degree it is possible to track it throughout the year. Figure 13 illustrates a solar tracking system with two degrees of freedom.

Figure 13 - Solar module with two degrees of tracking freedom
Source: Aton Tecnosol (2015)

3.2. 2Controller

Once the sun's trajectory has been traced and it is possible to determine the variations in the angle of incidence, it is possible to control the angle of the photovoltaic modules in an open loop without the need for sensors. The control can be done from a microcontrolled circuit, using a PIC or an Ardumo, which is easy to programme. Programming would be limited to sending an output signal to a motor at predetermined times corresponding to the time of the change in the angle of solar incidence. As the controller only needs pulses at programmed intervals, most of the time the control circuit would be in *sleep* mode, which means it would have low power consumption. To programme the controller, an RTC *(Real Time Clock)* is used where pulses can be activated by programming hours, days, months and years, making it inoperative when there is no need to send a signal to the output. This programming can be done based on a survey of how long it takes the sun to change its angle and how many degrees this variation is. This means that the gain that the board gets in terms of energy efficiency when its angle follows the sun's trajectory is not lost since the power supply to the controller circuit is very low. Figure 14 shows a model of an arduino uno board that is widely used for programming control circuits.

Figure 14 - Arduino Uno **Source: The** author

3.2.3Step Motor Drive

The photovoltaic module will be moved by a stepper motor that will be driven by the controller. There are two possibilities for driving the motor, one continuously where the motor would be switched on constantly and the other discreetly where the motor would only be driven at predetermined times, but with a view to low energy consumption, the discreet drive causes the motor to consume a very low level of energy and does not cause the energy generation system to suffer losses.

One of the most significant advantages of a stepper motor is its ability to be precisely controlled in an open-loop system. Open-loop control means that no *feedback* information about the position is required. This type of control eliminates the need for expensive detection systems such as optical encoders. Its position is known simply by controlling the input pulses. It is also very reliable, as there are no contact brushes in the motor. As a result, the life of the motor simply depends on the life of its bearing.

The simplest way to interface a unipolar stepper motor with Arduino is to use the ULN2003A *chip*. The ULN2003A contains seven *Darlington* transistors and is like seven TIP120 transistors all in one package. The ULN2003A supports up to 500 mA per channel and has an internal voltage drop of around IV when switched on. It also contains diodes for attenuating voltage transients when driving inductive loads. To control the stepper motor, voltage must be applied to each of the coils in a specific sequence. Figure 15 below shows the Ardumo connection for the stepper motor using a ULN2003A *chip*.

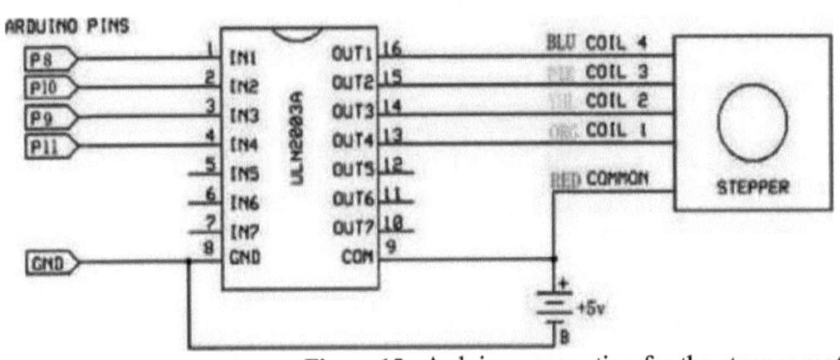

Figure 15 - Arduino connection for the stepper motor
Source: NADIEL (2014)

Commercially, you can buy the *kit* containing the ardumo board together with the stepper motor as shown in Figure 16.

Figure 16: Arduino and motor kit
Source: NADIEL (2014)

3.3 ENERGY GENERATION

Thinking about the energy generation of a photovoltaic system, an important item for energy efficiency is the choice of photovoltaic panel, as there are now different types of solar panels for use in companies, homes and solar power plants.

The energy efficiency of the photovoltaic panel is associated with the percentage of the sun's energy that reaches the surface of the photovoltaic panel and is transformed into electrical energy for our consumption. Other factors are also related to the panel's efficiency, such as:

- The more power per square metre your system will generate;
- The panel is smaller for the same energy production;
- Operate as closely as possible to an ideal generator, unaffected by temperature and weather variations.

3.3.1 Photovoltaic panel

Thinking about energy efficiency, cost and other factors related to energy generation, panels made from silicon (Si) are very viable for a photovoltaic energy generation project. According to Portal Solar (2015), around 80 per cent of photovoltaic panels in the world today are based on some variation of silicon. In 2014, around 85 per cent of all photovoltaic solar energy systems installed in homes and businesses worldwide used some form of silicon (Si)-based technology. Figure 17 shows the energy efficiency of different types of materials used in photovoltaic panels. Plates made from silicon have a lower efficiency than plates made from multi-junction cells, which are not used in residential systems. Although organic photovoltaic (OPV) cells also have higher energy efficiency than silicon panels, they are not widely used commercially, as today few companies have managed to take the production of OPV cells to an industrial scale. In Brazil there is CSEM Brasil, in Belo Horizonte, which is developing this production with mainly Swiss technology.

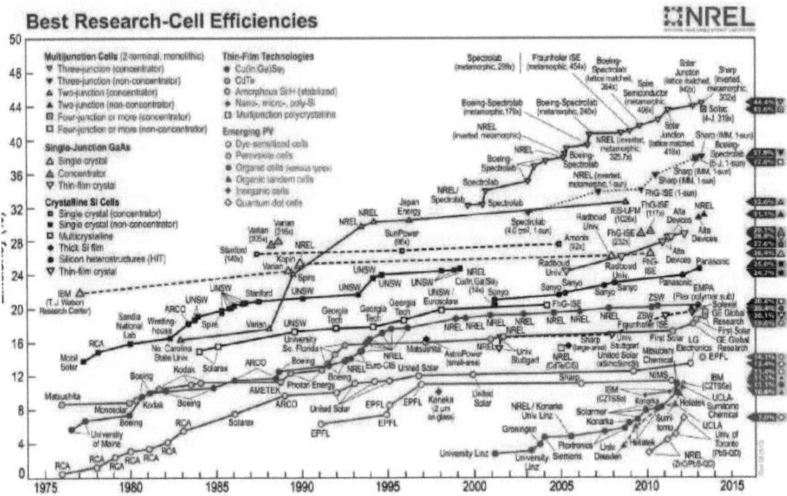

Figure 17 - Efficiency of solar cells
Source: NREL (2015)

3.3. 2Energy efficiency

In order to analyse the feasibility of automating the solar energy collection system, it is necessary to compare the energy efficiency of a fixed system with one degree of freedom and one with two degrees of freedom. A plate with one degree of freedom tracks the sun by changing its angle depending on the day, i.e. it changes its angle during the day following the sun from sunrise to sunset. With two degrees of freedom, the plate can follow the sun's trajectory during the different positions of the day as well as each season of the year, thus changing both the solar altitude angle and the azimuthal angle.

The control system is the same for both one and two degrees of freedom, because as it only sends electrical signals at predetermined times, the change would only be in the time taken to send the electrical signal from one degree of freedom to the other.

In order to take the gain from one system to another as a basis, it is possible to observe a study carried out on the automation of this system, where the power of the photovoltaic panels installed in a fixed way and with different degrees of freedom was measured. Figure 18 shows the efficiency of each.

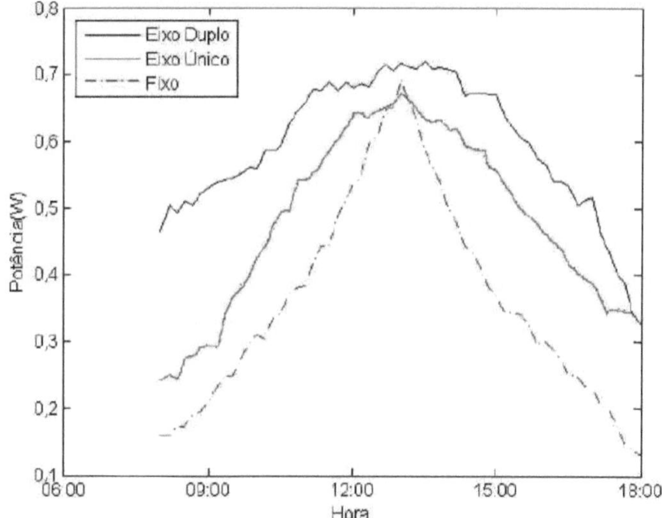

Figure 18 - Efficiency comparison **Source:** TREVELIN (2014)

Although this comparison was made by collecting the power values delivered only over a period of one day, the system using two degrees of ubiquity proves to be efficient, obtaining a gain of almost 40 per cent of the power delivered in a fixed system. According to Trevelin (2014), the power consumed by the controller circuit and the motor is less than 15% of the power received, which does not influence the system's gain.

Taking into account that the proposed system would not require the use of sensors, only a control to send signals to the motor, the power consumed would be even lower, since the electric current consumed would be in the micro Ampère *(p* A) range, which would mean that the power delivered by the board would have an almost negligible loss due to the controller circuit and the motor drive.

3.3.3 Solar inverter

Regardless of whether the system for generating energy using photovoltaic panels is made up of fixed or angle-shifted panels, the system must contain a solar inverter, which is the heart of your photovoltaic solar energy system. The purpose of the inverter is to convert the electrical energy generated by the panels from direct current (DC) to alternating current (AC). It also works as a system security system and measures the energy produced by the solar panels.

In homes, the solar inverter is typically installed near the light box, in a place sheltered from the sun, heat and water. The most widely used type of solar inverter is known as a *"grid tie inverter"*. These are the inverters used to connect the photovoltaic solar energy system to the electricity grid. The power (kW) required for this type of inverter depends on the power demand of the photovoltaic panels installed in the energy generation system, but it is also important to note whether the project foresees an increase in demand, because if this is foreseen, the inverter must be sized to support the increase in demand. The expression *grid-tie* in Portuguese means: connected to the grid. The *grid-tie* inverter is the inverter used to connect a photovoltaic system without batteries to the grid of your home or business. They are designed to quickly disconnect from the electricity grid if it goes down. Figure 19 shows the *grid-tie* inverter (Portal Solar, 2015).

Figure 19 - Grid-tie inverter **Source:** Energia Pura (2015)

As for the installation of the inverters, they should be fixed to a wall, or in a structure specifically built for the purpose, in an airy location that facilitates maintenance and control. The connection box with SPD and AC and DC switches must be installed close to the inverter. In some countries it is required that the disconnect switch be accessible to the power distribution operator so that he can switch it off if necessary, as a complementary safety measure.

The integration of the photovoltaic energy generation system with the electricity grid system supplied by the utility company is done in parallel. Often the energy generated by the photovoltaic system that is not consumed is returned to the electricity grid, which generates credits in the form of a discount on the energy bill. Figure 20 shows the connection diagram for the photovoltaic system and the power utility connected to the residential electricity grid.

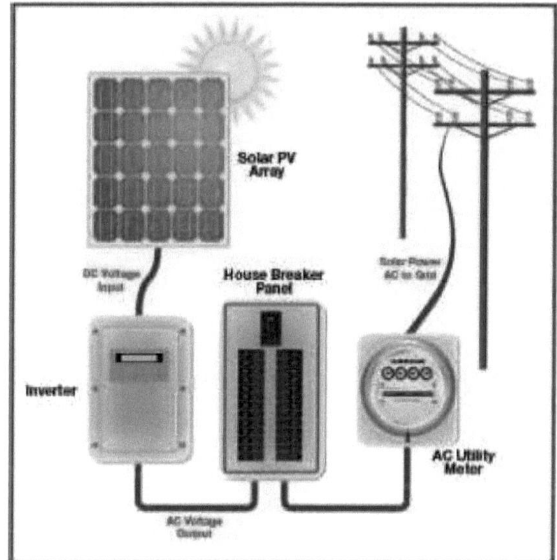

Figure 20 - Energy system connection diagram **Source:** Altemate Energy Company (2015)

3.4 WORKING PROTOTYPE

Although the aim of the proposal is not to design the project, prototypes of different blocks of the system were assembled to test them individually and thus verify the feasibility of its design, checking for possible defects and improvements to be made.

3.4.1 Variation in the angle of the plates

Firstly, on 11 August 15, with a temperature of 25°C and no cloud cover, two photovoltaic panels were placed side by side for a period of 6 hours. One of the panels was installed at a fixed angle of 25° to geographic north and the other was installed so that every hour its angle was changed following the path of the sun. This experiment was carried out to demonstrate only the variation in the angle of the photovoltaic panels throughout the day.

This experiment shows that discrete control is viable since the position of the sun does not change rapidly, which would require continuous control. Figures 21, 22, 23, 24 and 25 show the different positions of the PVs during the test periods. The technical information on the PVs used can be analysed on their *datasheets* in Appendix A.

Figure 21 - Sun position 09:50h **Source:** The author

Figure 22 - Positioning of the Sun 10:50 a.m. **Source:** The author

Figure 23 - Positioning of the Sun at 1:50 a.m. **Source: The** author

Figure 24 - Positioning of the Sun 12:50 pm
Source: The author

Figure 25 - Positioning of the Sun 1.50pm **Source: The** author

CHAPTER 4

RESULTS AND DISCUSSIONS

The proposed model is efficient in terms of energy generation, since, as previously shown, it is possible to obtain a gain in energy generation by changing the angle at which the sun's rays are captured by a photovoltaic plate. Although there is a circuit coupled to the board, energy consumption is irrelevant, as its power supply is around //A and is inoperative most of the time. The motor also has a low power consumption, as it will only receive a pulse at predetermined times, also remaining inoperative most of the time.

With regard to the additional materials that the proposed system needs to obtain for its implementation, this basically includes a board that controls the system by sending electrical signals to the motor, which can be an Ardumo Uno or a PIC, and also a stepper motor that receives the signals coming from the board. In the case of a system with two degrees of freedom, two stepper motors would be needed, one for each degree of freedom. You would also need a whole mechanical system that would be coupled to the motor and the PV so that it could rotate. All this additional material could cost between 6 and 7 per cent of the installation price of a fixed-plate system, which pays for itself over the years due to the energy efficiency gains from this system. Table 3 shows an estimate of the cost of implementing this system. The photovoltaic *kit* includes photovoltaic panels, *a grid-tie* solar inverter, *a* fixing structure, special cabling for direct current and special connectors. The photovoltaic *kit* is used regardless of whether the energy generation system is fixed or mobile with control of the angle of the panels. For the system with automation control, the Ardumo *kit*, the stepper motor and the metal structures interconnecting the motor shaft to the plates so that they can be rotated to change their angle are included. Considering the two degrees of freedom of the proposed control, the table shows the values for two Ardumo *kits*, two stepper motors and two three-metre steel bars for the metal structure.

Table 2 - Cost of an automatic solar tracking system

Material	Value	
photovoltaic *kit*	R$	12.500,00
Arduino *kit*	R$	196,00
Stepper motor	R$	480,00
Metal structure	R$	98,00

Source: The author

Analysing the systems for automatically capturing sunlight on existing photovoltaic panels, the proposed model has different characteristics and could be more efficient in terms of the power delivered by the panels. The models used as a basis used different types of sensors to capture changes in the sun's trajectory, so automating the circuit may result in greater energy consumption, i.e. discounting the energy produced with the energy consumed by the control circuit, the efficiency may not be as significant.

Mapping the solar trajectory shows the variation in the angle of movement throughout the year. This shows that a photovoltaic plate will not capture the same amount of sunlight at all times of the year. With this mapping it was possible to establish at which times this variation occurs, and this becomes a key element in the development of a control circuit for altering the angle of the PV.

One of the main advantages and disadvantages of a solar energy collection system that follows the path of the sun is the increase in energy efficiency, as studies have shown that an automatic, independent solar tracking system made up of one or two degrees of ubiquity has an advantage over a fixed system, as can be seen in Table 4.

Table 3 - Percentage increase in average power

	Single Axis	Fixed
Double axle	**18,8%**	**39,2%**
Single Axis	**-**	**25%**

Source: Trevelin (2014)

The disadvantage of this project is the possible maintenance that both the electronic and mechanical systems may need. Depending on the weather conditions, the photovoltaic plate may be able to control the angle, but not have an increase in energy generation due to a lack of solar incidence, or due to light dispersion due to clouds, so the angle adjusted by the control circuit may not be ideal for capturing the greatest incidence of solar rays on the photovoltaic plate.

CHAPTER 5

CONCLUSION

Within the existing projects and the proposal presented, it is possible to obtain various solutions that help photovoltaic energy systems to increase their energy efficiency. Some solutions feature continuous systems, most of which are more complex and seek to minimise variations in the sun, but have the disadvantage of greater energy consumption due to their control board, while others are simpler with discrete control, where the motor is activated at predetermined times, but which consume less energy and have a greater gain.

The cost of implementing this project depends on the energy demand to be generated, as this factor implies the number of photovoltaic panels used and their dimensions. However, for the automation system, a single control board can control several photovoltaic panels and a single motor can rotate more than one panel depending on its weight, i.e. it is possible to implement a mechanical system in which the panels are interconnected on the shaft of the same motor, not requiring the use of several motors, thus reducing costs for its implementation.

The proposal to automate the collection of solar rays for photovoltaic systems presented the characteristics, methods, materials and data that show how it is possible to implement this system and thus have better energy efficiency at all times of the day and year.

As a future direction, this proposal aims to design the project, pointing out improvements within the existing systems and putting into practice the proposal to automate the capture of solar rays in order to increase their energy efficiency in relation to photovoltaic energy generation systems with fixed plates.

CHAPTER 6

REFERENCES

ALMAKS, Laser Solution. **Laser Scribing.** Available at http://www.almakslasersolution.com/laser-scribmg/ - Accessed on 12/06/15

ALTERNATE ENERGY COMPANY. **Grid-Tie Systems Residential or Commercial**
Available at: http://altemateenergycompany.com/grid-tie-systems - Accessed on 03/08/15

ATON, Tecnosol. **Supports with automatic mobile tracker.** Available at http://www.atontecnosol.com.br/metalurgicaSuporte/movelAutomatico - Accessed on 22/07/15

CSEM, Brazil. **Organic Electronics.** Available at: http://www.csembrasil.com.br/p/eletronica organica - Accessed on 12/06/15

DE ALMEIDA SANTOS, Vinícius Puga. Stepper Motor. Available at: http://minilink.es/3fgz - Accessed on 01/09/15

DEL TORO, Vicent. **Fundamentals of Electrical Machines.** Rio de Janeiro: Editora Livros Técnicos e Científicos Editora S.A, 1994.

DUFFIE, John A.; BECKMAN, William A. **Solar Engineering of Thermal processes.** New York etc.: Wiley, 1980.

ELECTRONICS. Installation of Photovoltaic Solar Systems. Available at: http://www.electronica-pt.com/instalacao-sistema-fotovoltaico - Accessed 24/06/15

EVERYDAY ELECTRONICS. **Photo Resistor, LDR.** Available at: http://eletronicadiaria.blogspot.com.br/p/foto-resistor-ldr.html - Accessed on 19/06/15

PURE ENERGY. **Grid Tie Inverter SMA 10 kW Sunny Tripower.** Available at: https://www.energiapura.com/content/inversor-grid-tie-sma-10-kw-sunny-tripower - Accessed on 03/08/15

ENERGY INFORMATIVE, the homeownefs guide to solar panels. **Best Thin Film Solar Panels - Amorphous, Cadmium Telluride or CIGS?** Available at http://energyinformative.org/best-thin-film-solar-panels-amorphous-cadmium-telluride-cigs/ - Accessed on 12/06/15

INEO, National Institute of Organic Electronics. **First all-carbon solar cell created.** Available at: http://www.ifsc.usp.br/~ineo/news/index.php7pos id=212 - Accessed on 12/06/15

LAMBERTS, Roberto. **Orientation and Solar Diagram.** Available at: http://goo.gl/lN8jXb - Accessed 08/04/15

MILEA. P.L., OLTU. O, DRAGULINESCU. M, DASCALU. M. **Optimising Solar Panei Energetic Efficiency Using an Automatic Tracking Microdetector.**

Electronics, elecommunications and Information Technology Faculty "Politehnica" University of Bucharest luliu Maniu 1-3, Bucharest ROMANIA, 2007

MOLGARO. R.J. **Introduction to Photovoltaic Solar Energy.** Available at: http://pt.sHdeshare.net/RobsonJosuMolgaro/introduo-a-energia-solar-fotovoltaica - Accessed on 08/02/2015.

NADIEL, Comercio. **Controlling a 5v Stepper Motor with Arduino.** Available at: http://www.nadielcomercio.com.br/blog/2014/05/13/controlando-um-motor-de-passo-5v- com-arduino/ - Accessed 24/07/15

NREL. **NREL Sets New World Record with Two-Junction Solar Cell.** Available at: http://energyinformative.org/nrel-efficiency-record-two-junction-solar-cell - Accessed on 27/07/15

SOLAR PORTAL. **Types of photovoltaic solar panel.** Available at: http://www.portalsolar.com.br/tipos-de-painel-solar-fotovoltaico.html - Accessed on 27/07/15

RIBEIRO. S. C, PRADO. P. P. L, GONÇALVES. J. B. **Design and Development of a Solar Tracker for Photovoltaic Panels.** Symposium on Excellence in Management and Technology. 2012.

ROSÁRIO, João Maurício. **Principle of Mechatronics.** 6ª Ed São Paulo: Editora Pearson Education do Brasil, 2011.

SIEMENS. **Direct Current Motors.** 2006

TOPSKY, Electronics Technology(HK)Co, LTD. Available at http://www.topsky-tech.com/156mm-monocrystanine-solar-cell-supplier.html - Accessed on 11/06/15

TORRES, Gabriel. **Electronics for Autodidacts, Students and Technicians.** Rio de Janeiro Nova Terra Editora e Distribuidora LTDA.

TREVELIN, FeHpe Camargo. **Comparative study of solar tracking methods applied to photovoltaic systems.** 2014

VILLALVA, Marcelo G.; GAZOLI, Jonas R. **Energia Solar Fotovoltaica:** Conceitos e Aplicações. 1 ed. São Paulo: Editora Érica, 2012

VILLULLAS, H. Mercedes; TICIANELLI, Edson A.; GONZALEZ, Ernesto R. Fuel cells: Hmpa energy from renewable sources. **Redes,** 2011.

VIRIDIAN, Ecotechnology. **Photovoltaic Solar Energy** available at http://www.viridian.com.br/tecnologia/energia+solar+photovoltaic/4 - Accessed on 10/06/15

ANNEXES

ANNEX A - DATASHEET PHOTOVOLTAIC BOARD MODEL KC 130 TM

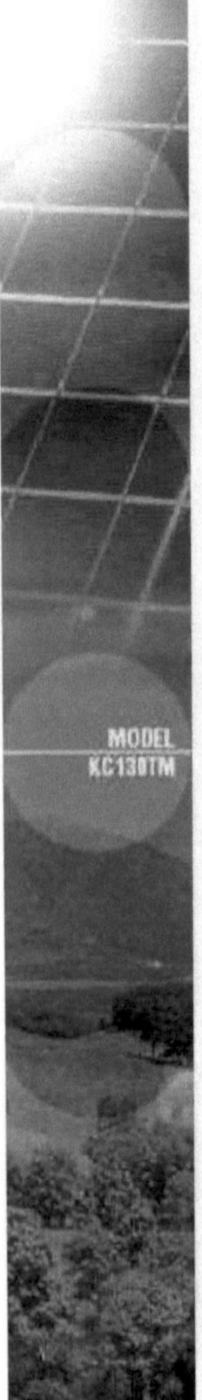

THE NEW VALUE FRONTIER

KC130TM

HIGH EFFICIENCY MULTICRYSTAL PHOTOVOLTAIC MODULE

HIGHLIGHTS OF KYOCERA PHOTOVOLTAIC MODULES

Kyocera's advanced cell processing technology and automated production facilities produce a highly efficient multicrystal photovoltaic module.
The conversion efficiency of the Kyocera solar cell is over 16%.
These cells are encapsulated between a tempered glass cover and a pottant with back sheet to provide efficient protection from the severest environmental conditions.
The entire laminate is installed in an anodized aluminum frame to provide structural strength and ease of installation.

APPLICATIONS

- Microwave / Radio repeater stations
- Electrification of villages in remote areas
- Medical facilities in rural areas
- Power source for summer vacation homes
- Emergency communication systems
- Water quality and environmental data monitoring systems
- Navigation lighthouses, and ocean buoys
- Pumping systems for irrigation, rural water supplies and livestock watering
- Aviation obstruction lights
- Cathodic protection systems
- Desalination systems
- Railroad signals
- etc.

QUALIFICATIONS

- MODULE : UL 1703 certified
  Hazardous Locations Class I, Div 2, Groups A, B, C and D
- FACTORY : ISO9001 and ISO 14001

QUALITY ASSURANCE

Kyocera multicrystal photovoltaic modules have passed the following tests.

- Thermal cycling test ● Thermal shock test ● Thermal / Freezing and high humidity cycling test ● Electrical isolation test
- Hail impact test ● Mechanical, wind and twist loading test ● Salt mist test ● Light and water-exposure test ● Field exposure test

LIMITED WARRANTY

※1 year limited warranty on material and workmanship
※20 years limited warranty on power output: For detail, please refer to "category IV" in Warranty issued by Kyocera

(Long term output warranty shall warrant if PV Module(s) exhibits power output of less than 90% of the original minimum rated power specified at the time of sale within 10 years and less than 80% within 20 years after the date of sale to the Customer. The power output values shall be those measured under Kyocera's standard measurement conditions. Regarding the warranty conditions in detail, please refer to Warranty issued by Kyocera)

ELECTRICAL CHARACTERISTICS

Current-Voltage characteristics of Photovoltaic Module KC130TM at various cell temperatures

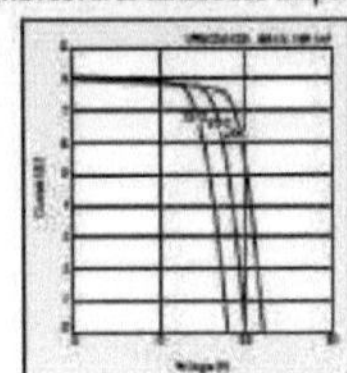

Current-Voltage characteristics of Photovoltaic Module KC130TM at various irradiance levels

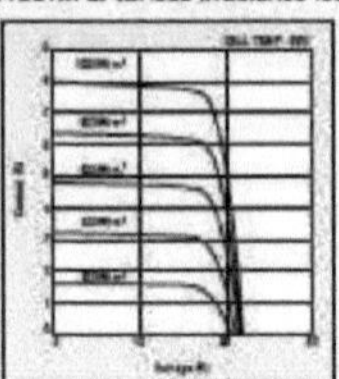

Physical Specifications

Unit: mm (in.)

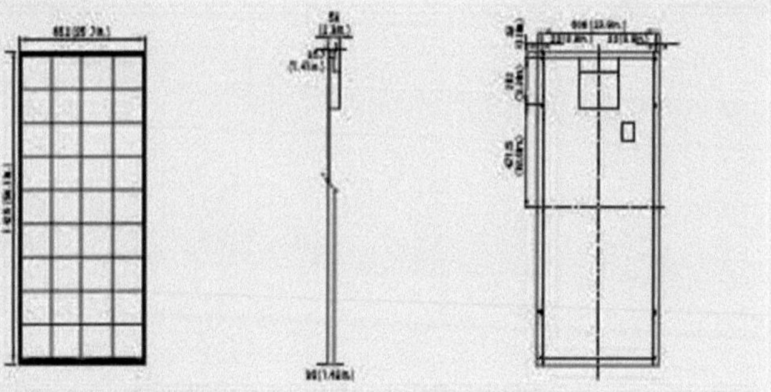

Specifications

■ Electrical Performance under Standard Test Conditions (*STC)	
Maximum Power (Pmax)	130W (+10%/−5%)
Maximum Power Voltage (Vmpp)	17.6V
Maximum Power Current (Impp)	7.39A
Open Circuit Voltage (Voc)	21.9V
Short Circuit Current (Isc)	8.02A
Max System Voltage	600V
Temperature Coefficient of Voc	-8.21×10^{-2} V/℃
Temperature Coefficient of Isc	3.18×10^{-3} A/℃

*STC: Irradiance 1000W/m², AM 1.5 spectrum, module temperature 25℃

■ Electrical Performance at 800W/m², NOCT, AM1.5	
Maximum Power (Pmax)	92W
Maximum Power Voltage (Vmpp)	15.6V
Maximum Power Current (Impp)	5.94A
Open Circuit Voltage (Voc)	19.9V
Short Circuit Current (Isc)	6.47A

NOCT (Nominal Operating Cell Temperature): [illegible]℃

■ Cells	
Number per Module	36

■ Module Characteristics	
Length × Width × Depth	[illegible]
Weight	11.9kg (26.6lbs.)

■ Junction Box Characteristics	
Length × Width × Depth	[illegible]
IP Code	IP65

■ Reduction of Efficiency under Low Irradiance	
Reduction	4.3%

Reduction of efficiency from an irradiance of 1000W/m² to 200W/m² (module temperature 25℃)

Please contact our office for further information

KYOCERA

KYOCERA Corporation

■ KYOCERA Corporation Headquarters
CORPORATE SOLAR ENERGY DIVISION
6 Takeda Tobadono-cho
Fushimi-ku, Kyoto
612-8501, Japan

● KYOCERA Solar, Inc.
Scottsdale, AZ, USA
http://www.kyocerasolar.com

● KYOCERA Solar do Brasil Ltda.
Recreio dos Bandeirantes, Rio de Janeiro, Brazil

● KYOCERA Solar Pty Ltd.
North Ryde
N.S.W., Australia
http://www.kyocerasolar.com.au/

● KYOCERA Fineceramics GmbH
Esslingen, Germany
http://www.kyocerasolar.de/

● KYOCERA Asia Pacific Pte. Ltd.
Central Plaza, Singapore

● KYOCERA Asia Pacific Ltd.
Tower 1 South Seas Centre, 75 Mody Road,
Tsimshatsui East, Kowloon, Hong Kong

● KYOCERA Asia Pacific Ltd. Taipei Office
Nanking West Road, Taipei, Taiwan

● KYOCERA(Tianjin) Sales & Trading Corporation
Chao Yang District, Beijing, China

Kyocera reserves the right to modify these specifications without notice

yes

I want morebooks!

Buy your books fast and straightforward online - at one of world's fastest growing online book stores! Environmentally sound due to Print-on-Demand technologies.

Buy your books online at

www.morebooks.shop

Kaufen Sie Ihre Bücher schnell und unkompliziert online – auf einer der am schnellsten wachsenden Buchhandelsplattformen weltweit! Dank Print-On-Demand umwelt- und ressourcenschonend produzi ert.

Bücher schneller online kaufen

www.morebooks.shop

info@omniscriptum.com
www.omniscriptum.com

Printed by Books on Demand GmbH, Norderstedt / Germany